Homemade Repellents

31 Organic Repellents and Natural Home Remedies to Get Rid of Bugs, Prevent Bug Bites, and Heal Bee Stings

© **Copyright 2016 by Daniel Beaumont - All rights reserved.**

This document is geared towards providing exact and reliable information in regards to the topic and issue covered. The publication is sold with the idea that the publisher is not required to render accounting, officially permitted, or otherwise, qualified services. If advice is necessary, legal or professional, a practiced individual in the profession should be ordered.

- From a Declaration of Principles which was accepted and approved equally by a Committee of the American Bar Association and a Committee of Publishers and Associations.

In no way is it legal to reproduce, duplicate, or transmit any part of this document in either electronic means or in printed format. Recording of this publication is strictly prohibited and any storage of this document is not allowed unless with written permission from the publisher. All rights reserved.

The information provided herein is stated to be truthful and consistent, in that any liability, in terms of inattention or otherwise, by any usage or abuse of any policies, processes, or directions contained within is the solitary and utter responsibility of the recipient reader. Under no circumstances will any legal responsibility or blame be held against the publisher for any reparation, damages, or monetary loss due to the information herein, either directly or indirectly.

Respective authors own all copyrights not held by the publisher.

The information herein is offered for informational purposes solely, and is universal as so. The presentation of the information is without contract or any type of guarantee assurance.

The trademarks that are used are without any consent, and the publication of the trademark is without permission or backing by the trademark owner. All trademarks and brands within this book are for clarifying purposes only and are the owned by the owners themselves, not affiliated with this document.

Table Of Contents

Introduction

Chapter 1: Natural versus Manmade Chemicals

Chapter 2: Insects and Your Health

Chapter 3: 31 Natural Repellents for Wasps, Termites, Ants, Mosquitoes, Roaches, Flies, Ticks, Spiders, Bed Bugs, Cloth Moths, Lizards, and Other Outdoor Pests

Chapter 4: Natural Ways to Bug-Proof Your Home

Conclusion

*****FREE BONUS*****

Introduction

I want to thank you and congratulate you for downloading **Homemade Repellents: 31 Organic Repellents and Natural Home Remedies to Get Rid of Bugs, Prevent Bug Bites, and Heal Bee Stings.**

This book contains proven steps and strategies on how to avoid using dangerous manmade chemical repellents, and to switch over to more effective and healthier natural repellents. We go outside into nature to get away from the indoor world we have created for ourselves. Basking in nature's God-given beauty is a fantastic way to ease the mind and live a healthier life, but bug bites and bee stings can make the outdoors a nuisance.

You have probably been solving this problem with bug repellents that you buy at your local grocery store or pharmacy. These repellents are made with dangerous chemicals that can harm your skin and endanger your health. By switching over to organic and natural ingredients, you will put yourself in better health while getting better protection from pests while outside.

Here's an inescapable fact: you will need a guide to the world of natural repellents. There is a lot of misinformation being spread out there, from natural repellents that are simply ineffective, to some that call for the use of dangerous chemicals. The goal is to rid you of the dangerous chemicals that we spray on ourselves when we go outside.

The knowledge found in this book will help you establish a list of go to ingredients when spending time with pests. By the time you complete this book, you will avoid the odor of chemical sprays and will further enjoy your time outside knowing you are protected with an all natural substitute.

If you do not develop your knowledge about natural repellents then you will forever be relying on dangerous chemicals found in common bug sprays. These repellents can cause skin rashes, nausea, and other unknown side effects. The long term effects of these sprays are unknown, but if the past is any indicator of the future, then these manmade sprays will continue to show new and pervasive health problems for generations to come.

It's time for you to make the switch from chemical to natural, enjoy the outdoors, and make your home a bug free haven. Make a change in your life today and you will forever be grateful. Break the chains of chemical sprays and step into the healthier world of natural repellents.

Chapter 1: Natural versus Manmade Chemicals

A revolution has taken place across America, and much it started just in the last decade. Americans are starting to realize that their lives are filled with dangerous chemicals. From the repellents sprayed on crops as they grow, to chemicals and preservatives in every product in the supermarket. Only now are we starting to see the dangerous effects of these chemicals. The overuse of strong chemicals and preservatives is part of the reason that Americans are so overweight, and their use correlates to rates of cancer and autism.

I am glad that you are taking the time to look at one aspect of your life, asking "Just how can I remove the dangerous chemicals from my bug repellents?"

This is one aspect of our life where manmade chemicals are completely unnecessary. There have been naturally forming bug repellents in use for thousands of years. Our ancestors did not live as long as we do today, but an educated guess would state that peoples before the use of chemicals would have less instances of cancer and other modern negative medical conditions.

The awareness of how manmade chemicals are bad for the body did not happen overnight. It took decades and thousands of needless deaths before people started to reanalyze the ingredients that they were putting into their bodies. The events that led to our newfound realization were tragic and took human lives, but our ignorance does not need to be our downfall anymore.

We know the incidence of the past, and we are prepared to move forward into the future and steer away from these dangerous chemicals. We must also be realistic about the limits of natural repellents and what they cannot do. There will always be a purpose for our chemically driven pesticides and repellents, but we should not have to deal with these in our everyday lives.

Going Natural

In our everyday life, natural repellents for bugs and other insects can provide far better protection. We can reapply ointments and sprays in a frequency far greater than manmade chemical sprays. These natural solutions are also cheaper to create and do not harm the environment. We can pack natural repellents in many different types of containers, and none of them need the use of aerosol. Not only are we saving our environment by not using manmade chemicals, but we are also preserving our atmosphere too.

The primary reason that I switched to natural repellents three years ago deals with my daughter. She has a fondness for nature that I also share. When we started going for hikes in the woods we were still using manmade repellents. As much as I love hiking and spending time with my daughter, nothing was quite as frustrating as going through the routine of taking out the bug spray and covering my daughter and myself. I would say to her "Don't breathe" and she would simply respond "Why not?" We had this little exchange a few times before I finally started to give some thought to her question – why was I afraid that she was breathing in these store bought repellents?

It only took a little bit of research to cement an idea that I had long already thought, children truly are wise beyond their years. My daughter was simply able to sense that the sprays and creams we had been using were in fact terrible for our health. They cling to the human skin for dozens of hours and are nearly impossible to wash off. This is by design to create a long lasting formula, but the cost is coating your body is chemicals that are nearly impossible to remove from your skin.

Switching to natural ingredients proved to be quite a learning curve, as I had to discover what natural formulas truly work and which ones don't (I have personally vetted the 31 solutions listed in Chapter Three). I saw in my daughter an improvement in how she felt during our hikes. She was able to enjoy the entire experience and was not left to dread the beginning of our walks when we'd typically spray ourselves down. I also found that the solutions we had come up with truly worked. They were proving to be just as effective as their chemical ridden counterparts, but now I could hike with my daughter and not have to worry about her safety in regards to her long term health.

While natural repellents are certainly the best way to protect you and your family, they do come with some limitations. Some chemical repellents do last significantly longer then natural solutions and there are some aspects to protecting your home that are simply be served better with manmade repellents. You should never use a chemical solution inside of your home, but since organic repellents fade faster than chemical solutions, the outside of your home is the one place where this might be worthwhile. Even still, you should not be spraying outside your home with manmade repellents unless you absolutely must.

Take note of your home's region; try our natural solutions listed in this book, and only turn to chemicals if all else fails.

The Dangers of Manmade Repellents

A can of bug spray is designed to thwart off nature's natural creatures, but in the process it hurts other aspects of nature – these chemicals can make their way into groundwater and drinking reservoirs. Although the particle counts are quite small, the fact remains that they taint water supplies for dozens of miles; water supplies that eventually put these harmful chemicals into our bodies. In this way manmade sprays are harming our entire existence. We may think that we are just applying our repellents when we are outside, but in fact their usage is trailing much further and is affecting us for months and years after their initial use.

The effects from early herbicides in the pesticides that were used in middle America are still being felt today. It the 1980s, as cancer rates for these middle states were going through the roof, it was theorized that modern pesticides were the cause. It would be nearly another two decades until this was proven correct, but the damage had already been done. Further, much of this land where these chemicals were first used have bodies of water that run South. The rates of cancer among people that lived near where this water flowed increased all through the 1980s, 90s, and even today.

The dangers from manmade chemicals come in their effectiveness, and the absence of knowledge about their long term effects. These chemicals are designed for a singular objective and they often come to market without companies doing the necessary research to find the long-term effects. Even if we think that the repellents we are using today are not harmful, it is possible that in twenty years this could very well change. The only way to be truly safe is to take the smart approach and switch to natural repellents; repellents that we know are effective and contain ingredients that won't cause long-term health complications.

Chapter 2: Insects and Your Health

No matter the size of the bug or insect, do not be fooled, nature's little creatures can wreck havoc on the human body. As you begin to veer away from chemical repellents and start looking at all natural solutions, it is important to remember that you must be willing to use a repellent of some type no matter the situation.

The damage that a small insect or bug can cause to the human body can be extreme, not because these creatures harm the body with natural venoms, but more because they carry diseases from far away locations. Make sure that you are always using some form of protection, and as you switch to natural repellents, be sure to use enough so that you are not getting bit. Below are just a couple of the harmful diseases that lurk inside the insects that surround us in nature.

Ticks - Lime Disease

Ticks are one of the most common pests in nature. They lurk in wooded areas and in tall grass field. They are small enough that they can latch onto the skin and begin to harvest blood without the victim ever being aware that they've been bit. Typically such a bite is discovered only after the skin surrounding the bite forms a red circular rash. The harm from ticks, like other insects, comes in the disease that it all too commonly carries, lime disease.

For many years lime disease was diagnosed as mono and such transmissions through tick bites would go unnoticed. Symptoms of lime disease start with fatigue, but it can ultimately be fatal if left untreated. In Chapter Three you will find a strong repellent against ticks, one that will protect you regardless if you are walking through their domain.

Mosquitoes – Zika Virus, West Nile, and Many Others

The zika virus is new in modern terms; identified just earlier this year the virus has extremely negative effects on pregnant women. It causes the yet to be born fetus inside a woman to have a shrunken skull by the time of birth. The reason for shrinkage caused by zika is not yet known, but what has been discovered is the method of transmission, mosquitoes.

Mosquitoes prey on the blood of mammals, and will bite several victims before they die. They then transmit diseases from one victim to another, and sometimes in the process new diseases are formed. This is how zika developed and the only way to prevent further transmission is to either eliminate all mosquitoes or protect yourself from mosquito bites.

Transmission of diseases through mosquito bites has long been a problem and zika is hardly the newest scare brought on by these insects. The West Nile virus was a concern for years, especially around parts of the Southern United States and Florida. Transmission, like zika, happens when a mosquito bites a victim and starts feeding on his or her blood. As many negatives as mosquitoes have, ultimately they are a crucial element of many ecosystems, and so the only way to truly protect yourself while maintain nature's balance is to avoid getting bit. Chapter Three features several methods of repelling mosquitoes and soothing the wounds they cause if you do happen to get bit.

Chapter 3: 31 Natural Repellents for Wasps, Termites, Ants, Mosquitoes, Roaches, Flies, Ticks, Spiders, Bed Bugs, Cloth Moths, Lizards, and Other Outdoor Pests

Below you will find the ingredients necessary to create natural protection against all sorts of insects, bugs, and pests. These natural ointments and sprays absolutely work, but sometimes can be less effective in small amounts. Bugs are attracted to you for a wide variety of reasons, from smell to the color of your clothing. If you are finding that you are still a point of attraction for nature's inhabitants, then try increasing the amount of the natural solution – you will never need to worry about using too much of the product as everything is perfectly safe for your body. For extra information about how to protect your home, please refer to Chapter Four.

Wasps and Bees: Three Solutions

Ingredients needed: Essential peppermint oil, water, and a spray bottle.

Cinnamon and a spray bottle filled with water.

Honey and whole garlic cloves.

Bees and wasps can ruin any day in the outdoors. These nasty critters will sting and attack with far less hesitation than nearly any other pest. At the same time, wasps and bees tend to create their nests around our homes, typically in doorways or small corners around doorframes. Our solution for preventive care for wasps and bees is two pronged:

One, if you are traveling into the forest for camping or to hike, I suggest you pack with you a spray bottle packed with essential peppermint oil and water. All you have to do is take an empty spray bottle, and mix in equal parts essential peppermint oil and water. Be sure to mix this solution well, and to mix before each additional application. You won't need much of this spray to thwart off wasps and bees, but you will need to apply the spray once every hour or so. Since the amount required is not large, you do not need a large spray bottle. A container of three-hundred milliliters should do the trick for a family of four for up to eight hours of protection.

Two, to stop wasps and bees from setting up around your home, all you will need is a little bit of cinnamon and a spray bottle filled with water. Since these insects like to build up high, you will need to sprinkle cinnamon in the affected spots, or where you believe they are likely to set up their nest. You then take the spray bottle and squirt a little bit of water onto the cinnamon – the goal here is to create just a little bit of adhesion.

The cinnamon should be reapplied once every two weeks to a month, and don't worry about the amount of cinnamon that you use. Just be sure to use an amount where you can get a faint smell of cinnamon while standing on the ground. Wasps and bees are far more perceptive of this smell than humans and even a small amount will keep wasps and bees away long after the smell has faded from your senses.

If you have already received a wasp or bee sting there is one natural remedy that I recommend to ease the pain and reduce the swelling. You will need crushed garlic and honey to make this remedy. First, take a few clothes of garlic and chop them up until they are very finely ground. Do not use garlic power as a substitute as this is too granular, and the added ingredients make it less effective than just regular garlic.

Take the chopped up cloves and mix them in honey, stirring to ensure an equal distribution of the garlic. Take this mixture and apply it to the affected area. You will want the honey and garlic to sit for about twenty minutes before you wash yourself off. The formula should be immediately effective in reducing pain, and swelling will be reduced within ninety minutes.

Termites: Four Solutions

Ingredients needed: Dehumidifier.

Essential orange oil and a hand drill.

Large flat piece of cardboard and water.

Extreme hot and cold temperatures, a sunny day.

Termites can slowly gather for months, wrecking havoc on your home. You might not even realize that you have a termite infestation until months after they have started chewing up your property. Be aware that termites almost always gather in moist areas with soft wood. Even wood that is hard will be appropriate for a termite if the air quality is too humid for a long a period of time. This softening of the wood allows termites to chew through strands and weaken the structural integrity of where the termites are located. The following methods will all work to either get rid of termite infestations or to prevent them, but be sure to be on the lookout for conditions that are suitable for termites – conditions that are attractive means it's just a matter of time before termites invade.

The lowest impact procedure you can conduct is simply to dry out the area where the termites are located. Usually you will need to combine this with one or more other methods to fully rid yourself of the infestation, but this is a good start and will reduce damage while other methods are also killing termites. Take a dehumidifier and place it where the infestation is taking place. Be sure to leave it running constantly and set the power to high.

You will want the air quality to be as dry as possible in hopes of draining moisture from the wood. I personally tried this method in my attic and found that within three or four days, new termites were not joining the existing home invaders. Every day, I would see more of them sitting on the outside of the wood in my attic, telling me that they were having difficulty chewing through the hard wood created with the dehumidifier.

A second method, and one that I combined with use of a dehumidifier, is essential orange oil. This product is very strong, and you will only need to use a few drops at a time. To use the product most effectively, drill several holes in the affected area. These holes do not have to be particularly deep, an inch or inch and a half will do. You will then want to put in droplets of the essential orange oil into the wood.

This should kill off any termites hiding in the location and will start to be effective within just a few hours of application. I found that I only had to drill a very few select holes in my attic and needed roughly three or four applications of the orange oil to rid my infestation. This, combined with the humidifier stopped the pests entirely and gave me my attic back.

If the infestation is not in your home, you will want to try this third method. Take a piece of cardboard, preferably from a large box and flatten so it is all on the same plane. You will then take this large piece of cardboard and soak it in a little bit of water. The idea is that termites move to the area of least resistance. They love cardboard and making it wet will be even more attractive to the termites. A few hours after setting the cardboard out, simply take the piece and bring it outside to either be destroyed, or you can 'fan' the termites out of the cardboard and place them back into nature, allowing you to use the same piece of cardboard again. I tried this method when I found a few termites in my shack and it certainly did work. The shack was quite old and the wood was only a few years from having to be replaced, so ultimately I did destroy the entire unit, however after rebuilding I would turn to this method first as it is the simplest and just about the fastest acting.

As a final method the sun works to clear termites out of any piece of wood. Simply take the affected piece of wood and place it in direct sunlight. Termites are very sensitive to temperature, and at about one hundred twenty degrees Fahrenheit they will die. This method also works if you live in a cold environment. Moving the wood outside will kill termites provided the weather is close to freezing. This method is not always practical as many times termites will infect pieces of wood that are cannot be moved from the foundation of your home, but in the right conditions, it can work very well.

Ants: Four Solutions

Ingredients needed: Lemon juice.

Essential peppermint oil, water, bowl, and hand cloth.

Pure ground cinnamon.

Boiling water and dish soap.

Ant infestations are extremely common, so common in fact that odds are at some point in your life you had an ant infestation. The key to working with ants is that they are guided by their sense of smell. By messing with an ant's perception, you can force them to wander off from the collective. You do not need to actually worry about killing individual ants as they tend to die in just a few days. All you need to focus on is ruining their sense of direction and forcing them away from coming into your home. There are a few methods of doing this, but conceptually they all work nearly the same.

For starters, if the infestation is coming from the ground, all you will need is some lemon juice. Do not dilute the juice with water or any other liquid. Just take the juice and spray a few squirts near where the ants appear to be coming in your home. The lemon juice will keep them from using that entry point. I used this when I had an infestation in my kitchen and it was immediately effective.

If the ants appear to mostly be coming in through the walls, then a solution of one part peppermint oil to three parts water will be your go to answer. Mix the peppermint oil with water in a small bowl and then take a cloth and dab it several times in the solution. Take the cloth and then spread it lightly on the wall where the ants appear to be entering, and along the path the ants are following. This method is better on your walls than lemon juice as it won't create any stains, but works in much the same way. It is not as effective as closing off areas for ants to get in, but it messes with their sense of direction in much the same way as lemon juice and will prevent ants from grouping in other locations of your home.

A third method, and one that you can use on both walls and on the floor, is to use a little bit of pure ground cinnamon spread liberally around where the ants are entering. I used this method when trying to deal with ants that were coming in through my home's walls. It certainly was effective, but it also left some difficult gunk on the sides of my walls. For this reason, try the peppermint first and only use cinnamon if the other methods are ineffective.

The first three methods are all for use inside your home, but if you manage to find the source the ants, then you may want to deal with the hive directly. You will need a heat source nearby and a large pot, but the following is a surefire method to clear out any ant colony. Simply take a pot of water and heat it till the water is boiling. Then, and after the pot has been removed from the heating element, add a teaspoon of dish soap per quart of water in the pot.

Mix the soap into the water until the water looks appropriately 'soapy', then take the pot and pour the boiling water down the ant hole. The water does not need to be exactly boiling, but you want it to be hot enough that it kills the ants on contact. The intention of the soap is to distort the ants' senses and to prevent them from regrouping.

Mosquitoes: Six solutions

Ingredients needed: Greek catnip oil.

Lemon eucalyptus oil, water, and a spray bottle.

Thyme oil, olive oil, and a bowl.

Essential cinnamon oil, water, and a spray bottle.

Crushed lavender.

Vinegar, bar soap, grater or knife.

For me and many others, mosquitoes are the most annoying insects of the outdoors. They carry a myriad of diseases, and even in the best case conditions, leave behind a bite mark that can itch for several days. I have tried many different ways to solve the problem of mosquito bites and preventative mosquito care. I would try each solution in order, with the last two solutions doubling as being effective treatment for existing mosquito bites as well as preventative care.

My go to answer for mosquito bites to apply a little bit of Greek catnip oil on the skin before going outside. This product is far more effective than most store bought spray repellents, and lasts for up to three hours. You do not need to apply much of the oil for it to show its effectiveness, so you will only need a small container. Most Greek catnip oil is sold in small containers and I found that one sixty milliliters container was enough to protect two people for around five eight hour excursions.

The scent of catnip oil might be a little uncomfortable for some people, although it doesn't seem to bother my daughter or myself, so if the smell offends try this solution instead: take a container of lemon eucalyptus oil and mix this with water in a concentration of one part lemon oil to ten parts water. You can make this mixture directly in a spray bottle and apply it to any part of the body. If you do not wish to spray directly on the skin, you can simply spray into the air and walk through the mist. The effectiveness of this mixture is good, but I found it to not work as well as the catnip. You will need to apply this spray once every hour or so to keep the mixture at maximum effectiveness.

As a third way to avoid mosquito bites, try thyme oil mixed with olive oil. This will create a powerful ointment that lasts for hours. To make the mixture, take a bowl and mix in one part thyme oil for every eight parts of olive oil. You will want to apply the ointment liberally to the skin and want it to soak into your pores before going outside. This ointment can be effective up to six hours.

Another way to avoid mosquito bites is with cinnamon oil and water. This mixture is great for creating a high quality spray that can be applied to clothing. It won't create any stains and is effective in warding off mosquitoes for two hours or more. Take cinnamon oil and mix it with water in a ratio of one parts oil to twenty parts water. Make sure to mix this solution very thoroughly before applying, and shake well for each additional application. Remember, oil and water don't mix and so you will have to momentarily make the bottle uniform by mixing it before each use.

Both effective for before mosquito bites and after, crushed lavender is one of the easiest ways to protective yourself. Simply buy lavender from your local store, usually craft stores carry it, and crush it in your hands. You do not need a mortar and pestle, lightly crunching in your hands will do just fine. To bring the crushed lavender with you outside, simply place it in your pockets. The smell of the lavender will go through your pants and keep away mosquitoes. I like this method, and at first it was my daughter's favorite, however the time that it is effective really depends on a lot of conditions.

You are relying on the smell of the lavender to cut through your pants, and depending on your own odor, the wind, and the material of your clothing, it could prove to be quite effective or barely ward off mosquitoes at all. Crushed lavender, as much as it might not be the best mosquito repellent, is a great way to deal with treating mosquito bites. Take the crushed lavender and rub it gently on any existing bites. The lavender will cool the 'heat' from the bite and should reduce some of the itchiness.

Lastly, vinegar with dry soap is a great way to ward of mosquitoes, but an even better one for treating bites. Any household vinegar will do, but for the soap you will want to find unused bar soap. Take the fresh bar soap and either use a grater or a knife to break off little scrapes of soap. Place this soap in about two ounces of vinegar. As a preventative measure, apply the resulting product as an ointment. It is quite effective and will last for about two hours. For after you have gotten a mosquito bite, take the mixture and rub it in firmly into the skin. You should feel the bar soap rubbing against your skin and soothing the bite. It will cause the bite to heal faster and reduce some of the itching.

Roaches: Two Solutions

Ingredients needed: Mint Essential oil, water, and a spray bottle.

Diatomaceous dirt.

Believe me when I say that I am intimately familiar with roaches. I grew up just outside of New York City and they are a regular occurrence in the area. They can be some of the most difficult bugs to get rid of due to their high tolerance for even manmade chemicals, and their large size makes killing one an issue in terms of cleanup. I have a couple of solutions to roaches – I have tried both and they work well, however even though these are natural solutions, the two methods listed contain some of the strongest ingredients in this book. Be careful not to get too much of either remedy on your skin, and wash thoroughly if this does happen.

My go to method is mint essential oil and a mix of water. You only need a very small amount of mint essential oil, and the water is there simply to dilute the strength of the mint and to increase the volume of what you will be spraying. Take the mint oil and put five drops into a spray bottle for every one cup of water. Make sure to mix the water and oil very thoroughly before spraying. Similarly to other oil based solutions that use a mixture of water, the oil and water will become separated very quickly, so quickly in fact that if you are spraying for a few minutes you will want to mix the bottle around every minute or so.

You can take this spray and apply it to places all around your home where you have been seeing cockroaches. For me, that means the basement and the bathrooms. I spray the solution behind my toilet bowls, in my sinks, and in my showers. I then apply the spray to corners around my basement. I find that the spray is effective for up a month, however cockroaches are fickle and their return depends on a variety of conditions. In cases where the spray does not appear to be working, make sure that you are mixing it well enough and apply it around twice a month.

This should not kill the cockroaches in your home, but it should cause enough of an annoyance for them that they wish to leave your home.

A second solution comes from diatomaceous dirt. Diatomaceous dirt is a naturally occurring soil and can be purchased for around twenty dollars. You will want to apply this dirt around the corners of rooms where you have cockroaches. I did not apply the dirt to most of my bathroom, but have tried it behind toilets and in the corners or spots that weren't likely to get wet. One application of the dirt is all that is necessary. You should find that it works better as a preventative measure than a way for actively dealing with roaches, but the single application means that it is something you only have to do once.

Flies: One Solution

Ingredients needed: Lavender oil, citronella oil, mineral water, and a spray bottle.

Flies are my least favorite outdoor nuisance. They are attracted to sweets and to cooking meats and are extremely difficult to get rid of. I only have one solution for dealing with flies, but you will only need one. This method works wonders and is effective for up to eight hours. I coat myself in a little bit of the spray right before I'm about to barbeque and have had little annoyance from flies since.

You will need lavender oil, citronella oil, and a little bit of mineral water. Mix one parts lavender oil with two parts citronella oil and a dash of water. The water should be pure, so don't use tap water and instead opt for anything bottled. You need literally just a few drops of water to create the spray so don't overdo it. The spray will need to be mixed before each application, but just a few squirts can cover an entire body. I find that I do not need to actually cover my own body, but just spray a little bit into the air and walk through it. The odor is also strong enough that it protects the area around me while I'm grilling, so don't feel the need to spray around the outside of your barbeque grill.

Ticks: One Solution

Ingredients needed: White vinegar, water, spray bottle, and optional lavender oil.

Ticks carry with them lime disease and other ailments. They suck on the blood of their victims and have caused thousands of deaths worldwide. You will want to be very careful of ticks when going hiking in the woods, particularly in areas with tall grass. Ticks are small and almost impossible to see unless you are looking for them. Usually the first warning sign of a tick is a red circle around where they have latched onto the skin. I found one very good method for dealing with ticks - a spray that will last an entire day of hiking. You will want to make sure that you spray the mixture directly on any exposed skin, and be liberal as you apply the mixture.

You will need distilled white vinegar, water, and an optional oil to lessen the strong smell of the vinegar. Take two cups of white vinegar and mix them with one cup of water. I mix this solution with essential lavender oil, and add just enough to cover the smell of the vinegar. Don't worry about how thoroughly mixed the solution is, as the vinegar will be strong throughout the spray. This spray is really only necessary when going hiking, and while I haven't' had a tick bite since, you will still want to check the exposed areas of your body after a long day of hiking.

Spiders: One Solution

Ingredients needed: Essential lemon oil, dishwashing soap, water, and a spray bottle.

You will only need this one method of dealing with spiders. This spray is meant for use around your house. Make sure to spray it in the corners of every room, on both the floor and ceiling. You will also want to spray it around any cracks and crevices that you find, as this is where spiders tend to hide themselves. The most common room of the house for spiders is the basement, and you will want to apply this spray about once a month. The cooling and warming months are particular times of the year when spider infestations become more common around my region. During these times of year I spray once every two weeks.

You will need essential lemon oil, dishwashing soap, and water to use this method. The lemon oil is very strong and you will only need five drops for every quart of water. As for the dishwashing soap, I put in about an ounce after I have mixed the lemon oil and water. This mixture will need to be thoroughly shaken before each application, but not too much of the actual spray is need. One or two shots in each location where spiders tend to rest should do the trick. As an additional note, spiders and other bugs are attracted to light, especially in the evening hours. Turn off the lights in your basement and you should attract less spiders overall.

Bed Bugs: Two Solutions

Ingredients needed: Soap, water, and sponge.

Grain alcohol, water, spray bottle, and optional essential lemon oil.

Bed bugs can infest more than just your mattresses. They love areas that offer a little bit of warmth and have lots of room for them to run around in. Aside from beds, the bugs can actually be quite attracted to desktops and laptops, as well as other electronic devices. There are two guaranteed solutions for bed bugs. If you have them in your own home, I suggest that you use the first method, however if you are on the road then you will want to go with the second method. You should never have to deal with bed bugs on the road, but it is not an ideal world and having the second method in your back pocket will assure you that you can get a great night's sleep no matter where you are.

The first method will be your best bet to firmly eradicate any bed bug infestation you might have. It's an old world method, but it does work extremely well. Simply heat up some water and mix in soap until you have a nice mixture that naturally has bubbles that form towards the top. Take a fresh clean sponge and dip it into the mixture. You will want to rub the sponge over your mattress several times. You do not want to get the mattress so wet that it cannot dry, but instead just get the surface a little bit wet. The real goal here is to spread the soap, and the water just acts as a conductor for spreading it evenly around the surface of the mattress. After you've done both sides, and the side strips to the mattress, take a second fresh sponge and wipe down all surfaces of the mattress. You want to get it at dry as possible. The whole process should take about forty five minutes to an hour, and I would do this in the morning so that you have time for the mattress to dry out and you can sleep on it that evening.

For computers, couches, and other bed bug infested items, or if you're just on the road, you will need grain alcohol and water. Mix one part grain alcohol with three parts of water, and stir well. Put the mixture into a spray bottle and be sure to apply to any item you think has bed bugs. For electronic items, you will want to make sure that the device is off and that you are spraying as far inside of the device as you can. For computer cases, this means dismantling the computer and coating the inside of the case in the solution. Do not turn on any electronic device for at least twenty-four hours after spraying to allow the device to dry. While I have not had any infestations in electronic devices, I have used this on a couch that I inherited, and have made the mixture several times when on the road. If the solution has an unpleasant odor, mix in a few drops of essential lemon oil and this will mitigate some of the smell of the alcohol.

Cloth Moths: Three Solutions

Ingredients needed: Dried lavender.

Cotton balls, essential lavender oil, and grain alcohol.

Freezer and plastic bags.

Cloth moths will destroy your clothes if given the chance. They will chew through wool in just a matter of days, and their infestation can go unnoticed for quite a while. This is especially true if you live in a climate where you store your winter clothes for half the year. I always make sure to use one of these methods for my winter clothes so that they do not develop holes from moths during the summer, spring, and fall months.

The first method I would use is dried lavender. This product can be purchased online in large quantities relatively inexpensively. Take the lavender and put it in some of the pockets of your clothing. You do not need to do this with each and every article, but space them out across a few pieces of clothing. You will need to replace the lavender every two months, and this gives you a good opportunity to see if moths have gotten into your closet. If you notice moths or any holes in your clothing, add more dried lavender and spread it closer together in your jackets. This is the method I use and I haven't seen a moth or a hole in any of my winter clothes since.

A second method that works just as effectively requires cotton balls, essential lavender oil, and grain alcohol. Take a bowl and mix in five drops of essential lavender oil with three ounces of grain alcohol. Mix this solution extremely well and then dab cotton balls in the bowl. The cotton balls should not be completely soaked, but they should be moist enough that you want them to dry for an hour or two before sticking them in your closet. I used to use this method but have stopped because the smell of alcohol is just a bit too strong, however it is effective.

As a last resort you can try bagging your clothing and freezing it. This requires a lot of freezer space and will only be useful for people in very particular situations. For example, if you live in the United Kingdom where moths can cause more harm to clothes than in the United States, then this might be your method of choice. You will absolutely not get any moths attacking your clothes – they simply will not have the opportunity. If nothing else seems to work, use this method, but be weary of the storage space that this will consume as you likely will not be able to store all of your winter clothes this way. You may want to use this method on treasured wools that you absolutely cannot afford to lose.

Lizards: Two Solutions

Ingredients needed: Garlic powder.

Egg shells.

Lizard tend to be more of a problem in hotter climates, but their presence can scare children and they tend to feast on beautiful plants that you would rather preserve. There are a couple of methods that work for repelling lizards, and I tried both of these while renting a home in Florida. Using either method I did not encounter any lizards, however the first method was better for my family since it did not create as much of an offense odor.

I would first try garlic power. You do not need to mix this with any other ingredients, and garlic power is cheap enough that you do not need to dilute it with water. Take the garlic power and put it around the outside of your property. You will want to put extra around your air-conditioning/heating unit, as this tends to be a place for where lizards like to congregate.

The second method is equally as inexpensive, but the smell can truly be unbearable, for both the lizards and for you. Take egg shells and crush them, then sprinkle them around your house. Note that in direct sunlight they smell far worse than when put in the shade. In either scenario they work effectively against lizards. They hate the smell and will want to stay away from your property.

Cicadas: One Solution

Ingredients needed: Table salt.

Cicadas won't appear every summer but when they do, they make their presence known though their loud signature mating call. You won't be able to get rid of the noise entirely, but you will be able to get them away from your property very easily and very inexpensively. To do this, take table salt and sprinkle some on the outside of your home. You will just want a few grains in any one spot, and you will want to cover as large an area as you can. With the salt in place you can expect that cicadas will go elsewhere.

Gnats: Two Solutions

Ingredients needed: Essential lemon oil, water, and a spray bottle.

White vinegar, dish soap, and several small bowls.

Gnats are the bane of my daughter's existence. If there was ever a reason that she would not want to go outside, especially in the wet months, it is because of gnats. They come in droves and can annoy like no other pest can. They attack during the night and also during the day, so you'll want to use one method or the other depending on the situation.

The first method requires essential lemon oil and water. Mix the two ingredients together in a mixture of one part lemon oil to six parts water. This is one of our strongest solutions that we create using essential lemon oil, and as such I would not apply it to the skin. It is perfectly safe to do so, but it might cause some irritation and it simply isn't necessary to apply to the skin directly. Spray the solution in the air and walk through it for protection that will last four to six hours.

For the evening hours you will need white vinegar, dish soap, and several shallow bowls. Mix the vinegar with the dish soap and place it in the bowl to rest it outside your tent, or near to where you are sleeping. The odor should not be enough for you to smell, but it will certainly be enough to keep gnats away from your campsite. You will want several bowls of the mixture to place around your campsite to get the best coverage.

There is one final natural way of dealing with gnats, and it is a method that is just as effective as the other two. In the evening and when you are at your campsite, sitting by the fire will remove any gnats that are gathering around you. They absolutely despise the smoke and will immediately leave you alone. This solution is the least practical as a fire will not always be available to you, and you will neither be sleeping near enough to it, nor will it be burning the whole night, for it to be effective while you are sleeping. I do not count this as one of the thirty-one guaranteed methods, but I do mention it here because it absolutely works. Just be careful not to breathe in any smoke from the fire, and use the other methods when and where you can.

Bonus: Strengthen Any Spray or Ointment

Many of the repellents described in this chapter are stored in a spray bottles or applied to the skin as an ointment. I have had amazing success with the formulas listed above, but if you ever do feel that what you have simply isn't strong enough or just isn't cutting it, there is one ingredient you can add to make any spray or ointment stronger. Mix in a bit of citronella oil, making sure to only use the minimum amount you need to improve the strength of the spray or ointment. If you need it, the boost will be well worth it, but beware that you should not be using citronella too often. The smell can make some people sick and the oil has been known to irritate the skin when too much is applied. Still, it's a nice trick to have up your sleeve when the situation calls for it.

Chapter 4: Natural Ways to Bug-Proof Your Home

Chapter Three lists ways of repelling specific insects and bugs from your home, but there are some general tips that you can follow to further protect your property. First, take note of the climate that you are in and the most common insects and pests. In New York this means roaches and spiders, but in Florida it could mean lizards and moths. Use the prescribed methods from Chapter Three and you should be able to deal with your most common pests. Also keep note that most climates tend to cycle through the common bugs that infiltrate homes, so for several months of the year you may be able to avoid proofing your house against a particular type of insect or pest. The following methods work in just about every climate and range in price and required man hours.

Windows

If indoor pests are a problem, the first part of your home that you will want to look at is your windows. Many windows from the 80s and 90s were built with an old form of rubber to maintain the seal. As such, these can be easily infiltrated by insects and pests. The costs can be expensive, but sealing your windows will both raise the value of your home and protect you for many years to come.

To identify if your windows are indeed a trouble spot, take a survey of the outside area. There will be key structural signs that a window is not fulfilling its duty. Look for where the windows meet with the house to see if the plaster has any cracks or holes. These cracks do not need to be large enough for an insect to fit through – you are simply looking to determine the quality of the plaster work itself. If you find many cracks, you might want to have a professional look into the problem.

While not as effective, you can also monitor your heating and cooling bill and compare it to the average for your area. Most utility companies offer a guide to show you how much energy you use compared to the neighboring homes. If your bill is far above the average then it is possible that your windows have a small leak, or more likely the material around the window has small cracks.

Spraying The Cracks

There are small cracks in every home that go unnoticed for years. These cracks can be so small that you might even notice one but think to do nothing about it because you can't imagine an insect crawling in. Using the specific sprays from Chapter Three around these cracks is a great way to eliminate insects, but it is most effective on spiders. Spiders tend to lay eggs in extremely small cracks and a few squirts of one of your natural spider spray will kill the eggs before they have a chance to hatch. Do not worry about any cleaning anything within the cracks, the material is too small and to reach it would mean causing damage to your home.

To find these cracks can be long an arduous process, but it is worth it if you wish to fully eliminate spiders from your home. Cracks that house spiders tend to form in the archways of doors and around parts of the house that go through the most temperature changes in a given year; this means walls that sit in front of uncovered windows or your attic.

Salting

Regular table salt is a vastly underused insect and bug repellent. Take table salt and put it outside of your house to ward off several types of bugs. It is most effective on cicadas but will work on ants and other critters too. You do not need a strong concentration of salt on any one area, so just sprinkle it throughout the perimeter of your home and you should start seeing the effects. There is one case where you might not want to salt your property and that is if you have deer in your area. Deer will be attracted to the salt and they will seek the surfaces where it is located. Personally, I have deer in my area and I still salt the outside of my property. They are attracted to the salt but this just means that I lay down the salt a little bit further away from my home. It's worth the reduction in pests and insects and is both cheap and easy to do.

Conclusion

Thank you again for downloading **Homemade Repellents: 31 Organic Repellents and Natural Remedies to Get Rid of Bugs, Prevent Bug Bites, and Heal Bee Stings.**

I hope this book was able to help you make the switch from manmade chemical repellents to safe, natural remedies. You will need some time to prepare each and every solution that this book presents, but hopefully within a few months your bug protection bag is carrying nothing but these natural repellents. Feel free to experiment with these formulas and modify them as you see fit, but remember the key ingredients and the measures provided in each solution. Each method for repelling bugs requires each and every ingredient to be effective. If you do decide to change these formulas, my suggestion is to be additive, and not to take away from what is already in each formula.

The next step is to start crafting some of these natural recipes. It can be a fun, soothing activity, and you will feel good about making an alternative to a product that is both overpriced and dangerous for your health. I hope the switch to natural repellents is an enjoyable experience, and that the tips in this book will lead to a greater enjoyment of nature for you and your family. It's time for you to go outside and protect yourself from nature's pests with all natural ingredients, feel good about yourself while basking in the sun, and maintain a healthy lifestyle while staying bug free.

    Finally, if you enjoyed this book, please take the time to share your thoughts and post a review on Amazon. It'd be greatly appreciated!

    Thank you and good luck!

    - Daniel Beaumont

*****FREE BONUS*****

Thank you for staying tuned till the end! As a reward, I would like to give you 3 bonus materials: 2 PDF books and access to my favorite natural remedy database!

Just copy and paste the following URLs into your favorite internet browser:

BONUS 1: *The Creative Herbalist*

http://bit.ly/2bQSnGX

BONUS 2: *The Encyclopedia of Natural Remedies*

http://bit.ly/2c4uVro

BONUS 3: *Natural Insect and Pest Control Article Database*

http://bit.ly/2bJ9NGN